CORN

FARM TO MARKET

Jason Cooper

Rourke Publications, Inc.
Vero Beach, Florida 32964

12/98 JFɔnd12ᵒᵒ

Edited by Pamela J.P. Schroeder

PHOTO CREDITS
All photos © Lynn M. Stone

Library of Congress Cataloging-in-Publication Data
Cooper, Jason, 1942-
    Corn / by Jason Cooper.
        p.   cm. — (Farm to market)
    Summary: Examines the importance and different kinds of corn and explains how it is grown, processed, and shipped to the grocery store.
    ISBN 0-86625-620-2
    1. Corn—Juvenile literature. 2. Corn products—Juvenile literature. 3. Corn—Middle West—Juvenile literature. 4. Corn products—Middle West—Juvenile literature. [1. Corn.] I. Title. II. Series: Cooper, Jason, 1942- Farm to market.
SB191.M2C78234   1997
633.1'5—dc21   97-13056
                                                CIP
                                                AC

Printed in the USA

TABLE OF CONTENTS

CORN

We know corn best for the sweet taste of its kernels. Corn is more than a tasty, summertime treat. It is one of the most useful and important crops in the world!

Corn is a plant in the **grain** (GRAYN) family. Wheat is also a grain, along with plants such as rice, oats, and barley.

People around the world depend more on corn food products than any others, except those made from wheat.

Corn is one of the world's most important sources of food. This is dent corn, the most popular kind of corn in the United States.

KINDS OF CORN

Several thousand varieties of corn grow around the world. They all have roots, a spearlike stalk, leaves, and flowering parts. The corn cob, or ear, is one of the flowering parts.

The ear has rows of corn kernels. The kernels ripen under a wrapping of leaves that make up the corn **husk** (HUSK).

The best known groups of corn are sweet, pop, dent, and flint, or Indian. Most North American farmers grow dent corn.

Flint, or Indian, corn is colorful.
It's often used for fall decorations.

WHERE CORN GROWS

Corn grows wherever the **climate** (KLI mitt) is mild or warm. The world's leading corn-growing country, however, is the United States. It grows four of every 10 ears.

Much of the Midwestern United States is known as the Corn Belt. More than three-fourths of America's corn grows in the dark, rich soil of the Midwest. Illinois and Iowa are the largest corn growers.

Most Corn Belt corn is dent corn. Its stalks grow almost as high as a basketball rim.

Corn grows much taller than a man in the black, fertile soil of Corn Belt states.

PLANTING CORN

Corn plants grow from seeds planted by corn planting machines. One corn planter can plant as many as 24 rows of corn at once.

Most corn fields in the Midwest are prepared for planting by a field cultivator, pulled by a tractor. Farmers may treat the soil with a fertilizer. Fertilizers help the corn plants grow.

Planting in the Midwest begins in late April or May. By then the soil temperature is at least 50°F (12°C) and the seeds sprout quickly.

Young corn plants surround an old corn barn, or crib, in an Illinois field.

Like a slow-moving tornado on wheels, a combine mows down rows of corn. At the same time, the machine plucks and shells the ears.

Beef cattle fatten up on silage stored in huge silos (background).

GROWING CORN

Farmers in the Corn Belt watch the weather and the number of insects closely. Farmers can't change the weather, but they can hope for an average of 1 inch (2.6 centimeters) of rain each week. The growing season for their fields of corn is about 20 weeks.

Farmers deal with insect pests by spraying the fields with chemicals called **insecticides** (in SEK tuh sydz). Insecticides kill insects.

Corn Belt farmers grow as many as 28,000 corn plants in 1 acre (69,000 in 1 hectare). One acre produces about 150 bushels of corn.

One acre in America's Corn Belt can grow 28,000 corn plants. By July, each plant should have a ripening ear of corn like this one.

HARVESTING CORN

Midwestern farmers harvest their dent corn about six months after planting it. Farmers try to wait until the corn kernels dry out. Corn must be quite dry for storage.

An amazing machine called a combine harvests corn. The combine moves slowly through corn rows like a giant caterpillar. As it knocks down stalks, the combine plucks the ears. It also removes the corn husks and **shells** (SHELLZ), or removes, the kernels from the ears.

Illinois farmers check a row of dent corn to see if it is dry enough. If it is, they will combine the corn.

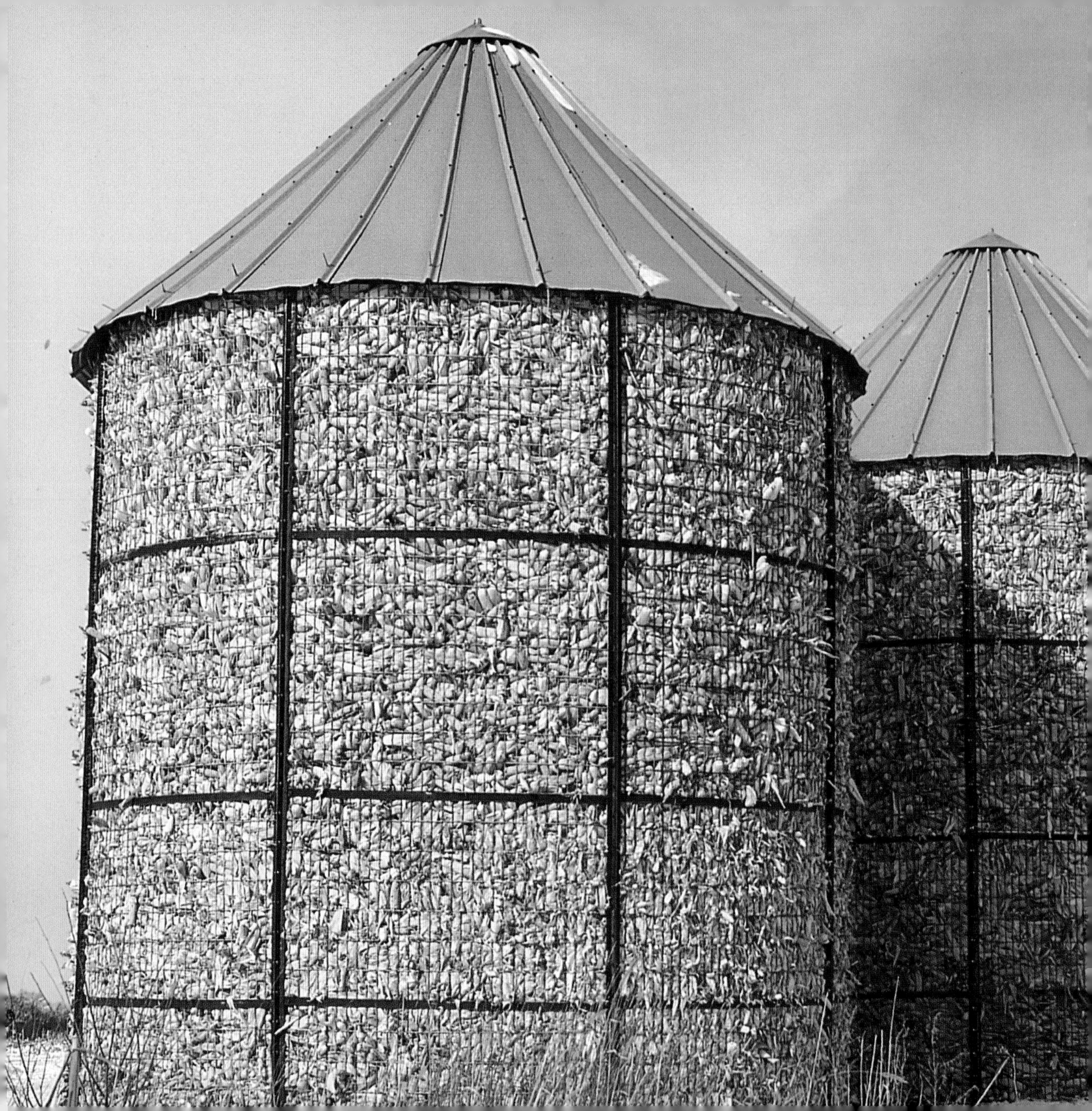

PROCESSING

Farmers store corn for animal food in huge silo-type bins. About half of America's corn, however, is **processed** (PRAH sest), or changed, into corn products.

Factories change corn through many steps. One step separates the different parts of corn kernels. Another step adds corn to other foods. Corn may be heated, soaked, or ground.

The many ways of processing corn result in many different corn products.

A farmer has stored corn for his hogs in huge wire bins.

CORN PRODUCTS

Some corn is used for **silage** (SI ledj). Corn silage is a ground mix of whole corn plants. It is stored in silos to feed livestock.

The corn processed by factories may be turned into cornflakes, corn meal, corn syrup, corn starch, or one of hundreds of other corn food products.

Corn is also used in some paper goods, building materials, medicines, fabrics, paints, and explosives. Ethanol and gasohol are engine fuels made with corn products.

Corn is processed into many food and non-food products. Sweet corn and a few corn food products are shown here.

CORN AS FOOD

Corn is an important food product. Sweet corn on the cob is delicious. More important, corn has some vitamins and a great deal of fiber. Fiber helps your body process food.

Corn also has a large amount of starch and some proteins. Starch helps give you energy.

More than 1,000 common food products, such as syrup, salad dressing, tamales, and tortillas, have corn in them.

Glossary

climate (KLI mitt) — the average weather (temperature, wind speed, moisture) of a place over a period of years

grain (GRAYN) — the "cereal" grasses, or the seed or fruit of cereal grass; corn, wheat, rice, and kin

husk (HUSK) — the leaves covering the kernels on an ear of corn

insecticide (in SEK tuh syd) — chemicals used to kill insect pests

processed (PRAH sest) — changed in form or moved from one place to another; prepared for market

shell (SHELL) — to remove corn kernels from the ear

silage (SI ledj) — ground-up corn plants used for livestock feed and stored in silos

INDEX